Earth in Balance

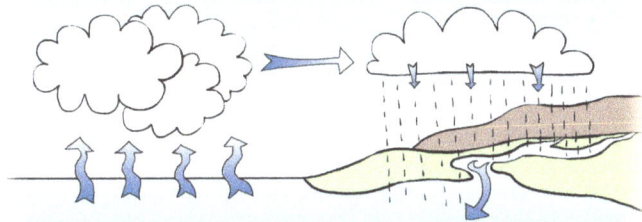

Illustrations: Janet Moneymaker
Design/Editing: Marjie Bassler

Earth in Balance
ISBN 978-1-953542-29-8

Published by Gravitas Publications Inc.
Imprint: Real Science-4-Kids
www.gravitaspublications.com
www.realscience4kids.com

RS4K

Earth has different parts that work together to keep Earth in **balance**.

Geosphere
The rocky part of Earth.

Atmosphere
All the air on Earth.

Biosphere
All the living things on Earth.

Hydrosphere
All the liquid water, snow and ice on Earth.

Magnetosphere
The magnetic field surrounding Earth.

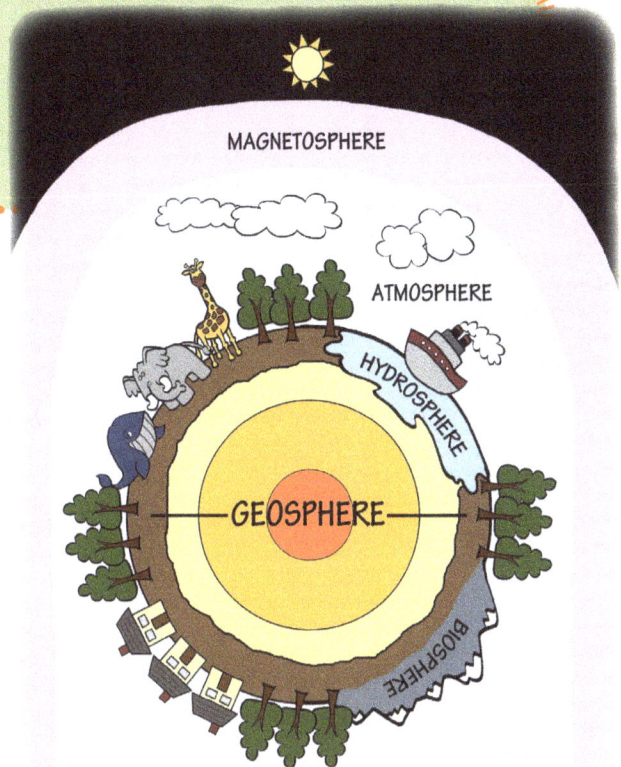

All of Earth's parts are needed
for life to exist on Earth.

Animals
would die too.

Rain and snow in the **hydrosphere** give plants and animals the water they need to live.

The rain and snow go into rivers and streams that flow into the oceans. Water from rivers, streams, and oceans goes back into clouds in the air where it can fall again as rain and snow.

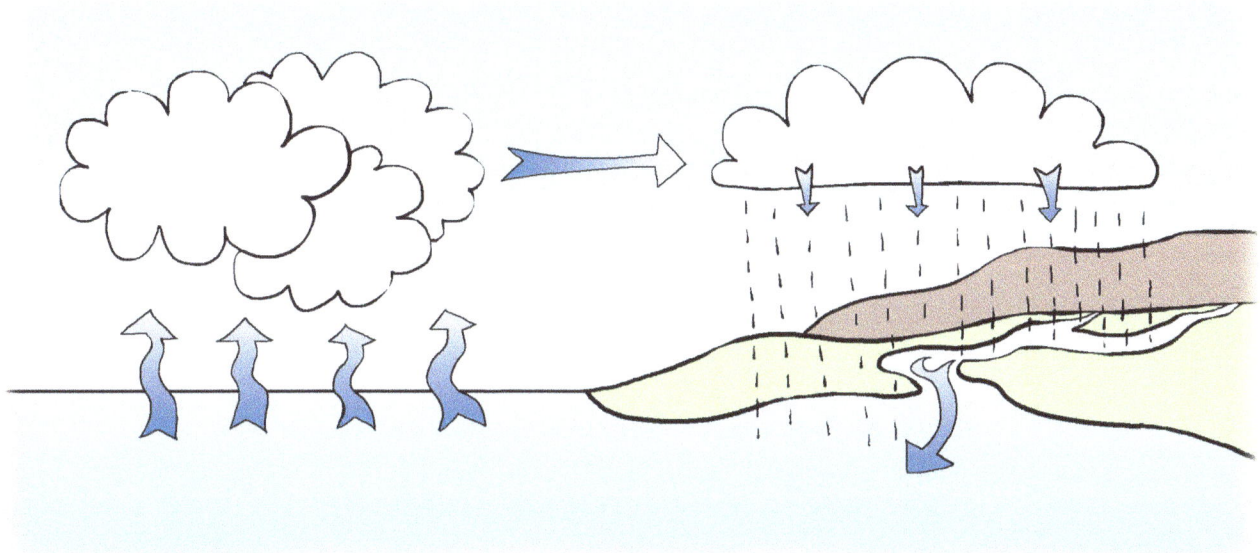

The air in the **atmosphere** has **atoms** and **molecules** that living things need.

Plants take **carbon dioxide** from the atmosphere and change it into **oxygen**. Animals breathe in oxygen and breathe out carbon dioxide.

ATOMS

I am a carbon atom.

I am an oxygen atom.

Atoms are tiny building blocks that can link together.

Atoms make up everything we touch, taste, smell, and see.

REVIEW

MOLECULES

We are a carbon dioxide molecule.

Molecules are made when **atoms link** together.

Just enough of the Sun's **energy** gets through the **magnetosphere** and the atmosphere for plants and animals to stay alive.

If we got too much sun, we would burn up.

That would be icky.

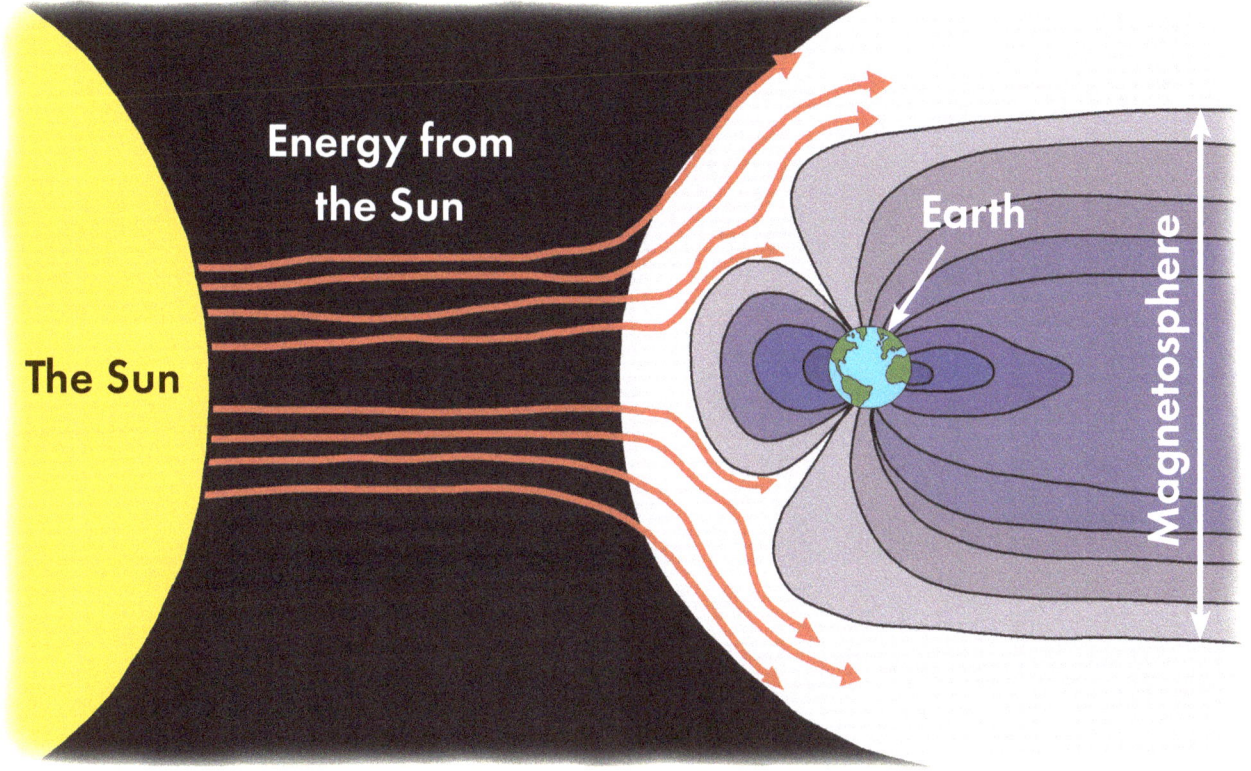

The Sun

Energy from the Sun

Earth

Magnetosphere

When all the parts of Earth are in balance, plants and animals can stay alive and healthy.

But it is possible for Earth to become unbalanced.

If there is too much carbon dioxide in the air and not enough plants to change it to oxygen, weather patterns can change. Too much carbon dioxide also makes oceans become less healthy for fish.

Would you like to be a fish?

Do they eat cheese?

Changing weather patterns can make some areas have too little rain and some areas have too much rain.

If too much ice melts, sea levels can rise. This leaves less land for plants and animals to live on.

We can help keep Earth in balance by cleaning the air and the water in lakes, rivers, and the oceans. We can choose to use products that are Earth friendly, use fewer plastics, and recycle. Everyone can help!

 Let's help!

 YES!

How to say science words

atom   (AA-tuhm)

atmosphere   (AT-muh-sfeer)

balance   (BAA-luhns)

biosphere   (BIY-uh-sfeer)

carbon dioxide   (KAHR-buhn diy-OCK-siyd)

energy   (EN-uhr-jee)

geosphere   (JEE-oh-sfeer)

hydrosphere   (HIY-droh-sfeer)

magnetic field   (mag-NE-tik FEELD)

magnetosphere   (mag-NEE-tuh-sfeer)

molecule   (MAH-luh-kyool)

oxygen   (OCK-sih-juhn)

science   (SIY-uhns)

What questions do you have about EARTH IN BALANCE?

Learn More Real Science!

Complete science curricula from Real Science-4-Kids

Focus On Series

Unit study for elementary and middle school levels

Chemistry
Biology
Physics
Geology
Astronomy

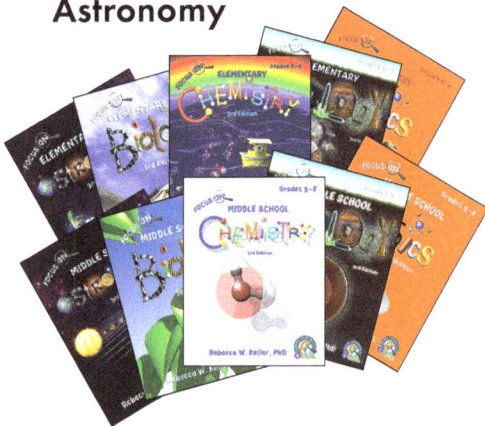

Exploring Science Series

Graded series for levels K–8. Each book contains 4 chapters of:

Chemistry
Biology
Physics
Geology
Astronomy

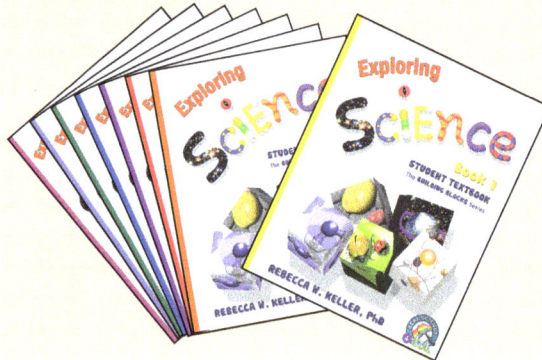

www.ingramcontent.com/pod-product-compliance
Lightning Source LLC
Chambersburg PA
CBHW040149200326
41520CB00028B/7542